Table of Contents

1 Introduction to Image Synthesis

Generative models form the bedrock of image synthesis. They differ fundamentally from other image manipulation techniques; instead of altering existing images, they create entirely new ones from scratch, often based on learned patterns or defined rules. Understanding this distinction is key to grasping the power and potential of this field. The ability to generate completely novel visual content opens up a vast array of creative and practical possibilities, spanning artistic expression to scientific data visualization. This fundamental concept will underpin all subsequent discussions in this book.

Early attempts at image generation were rudimentary, relying on simple algorithms and limited computational power. Consider, for example, the pioneering work with random dot patterns, where images were constructed by placing dots randomly on a canvas. While visually simplistic, these early experiments laid the groundwork for more sophisticated methods. They demonstrated the basic principle of building images algorithmically, a core concept that persists even in the most advanced modern techniques. This early work highlights the iterative nature of technological advancement, with each step building upon previous breakthroughs.

The shift from purely random methods to more structured approaches marked a significant step forward. Cellular automata, for instance, introduced the concept of local interactions generating global patterns. Simple rules governing the state of individual cells resulted in complex, emergent structures—a fascinating preview of the self-organizing nature often seen in complex systems. These early systems, though limited in their visual complexity, showcased the power of simple rules to generate surprising and intricate results, anticipating the power of later, more advanced systems.

Procedural generation emerged as a powerful technique, enabling the algorithmic creation of diverse and varied content. Unlike purely random approaches, procedural generation utilizes algorithms that follow specific rules to generate outputs. This allows for greater control and repeatability, making it particularly suitable for creating game environments or generating large datasets for training more complex models. It's a bridge between simple algorithms and the intricate architectures of modern deep learning techniques, allowing for customized content creation within specific parameters.

Moving beyond low-resolution methods, the use of vector graphics and their mathematical descriptions opened up new possibilities. Scalable Vector Graphics (SVG) became a cornerstone of digital design, enabling the creation of images that could be resized without loss of quality. This flexibility contrasts sharply with the limitations of pixel-based images, opening up greater possibilities for use across various scales and formats. The mathematical nature of SVGs also laid the groundwork for more advanced techniques based on geometric principles.

Fractals and L-systems represent a particularly elegant approach to image generation using mathematical formulas. Fractals, with their self-similar patterns at different scales, allow for the creation of stunningly intricate and complex imagery from relatively simple rules. L-systems, a type of formal grammar, provide a powerful tool for generating plant-like structures and other organic forms, demonstrating the use of mathematical structures for generating realistic and aesthetically pleasing images. These examples show how mathematical principles can generate complex and realistic looking images, demonstrating the intersection of mathematics and image synthesis.

The leap to deep learning marked a revolutionary shift in image synthesis. Deep learning's ability to learn complex patterns from large datasets opened up possibilities previously unimaginable. Generative Adversarial Networks (GANs), for example, utilize a competitive framework—a generator creating images and a discriminator judging their authenticity—to produce increasingly realistic results. These models have dramatically increased the quality and sophistication of synthetic images, becoming a central component of the field.

Variational Autoencoders (VAEs) offer a different approach, focusing on learning a latent representation of the data. This allows for generating images from a lower-dimensional space, providing a different perspective on the problem of generating synthetic images. This approach provides unique advantages, offering alternatives to GANs and opening up further avenues for research and development. The different methodologies highlight the diverse paths and directions within the field.

Diffusion models have gained prominence recently, offering a distinct advantage by modelling the process of gradually adding noise to an image and then reversing it to generate new samples. This approach has proven remarkably effective in producing high-quality, detailed images, showcasing a different paradigm for image synthesis. This shows the evolution of the field and highlights the ongoing development of novel techniques.

Autoregressive models represent another significant advancement, generating images pixel by pixel based on the preceding pixels. This sequential approach is computationally intensive but can yield highly detailed and coherent results. Their sequential nature offers a different perspective, demonstrating yet another path towards the goal of generating high-quality images. The diverse range of techniques highlights the depth and breadth of modern image synthesis.

Finally, Style Transfer techniques allow for the application of artistic styles to existing images. This transforms the visual appearance of an image without altering its core content, making it a powerful tool for artistic expression and image manipulation. It demonstrates the interplay between image generation and image manipulation, highlighting further possibilities within the field. The development of these diverse techniques has greatly expanded the possibilities available to artists and researchers.

1.1 Defining Generative Models

Generative models are the heart of synthetic image creation. Unlike traditional image processing methods that manipulate existing images, generative models create entirely new ones. They learn patterns from input data – be it images, text descriptions, or even audio – and use this learned knowledge to generate novel outputs. Think of it like teaching a computer to paint by showing it thousands of paintings; eventually, it can create its own, mimicking the styles and techniques it has observed. This learning process varies considerably depending on the specific model. Some models work by subtly modifying a random input until it resembles a real image from the training data. Others might build up an image piece-by-piece, predicting each pixel based on those that came before. The core concept, however, remains consistent: the model learns from examples and generates unseen outputs based on that knowledge. Understanding these fundamental principles is crucial for anyone working with image generation.

The choice of generative model greatly impacts the resulting image. For instance, models that build images pixel by pixel generally produce sharp and well-defined outputs, but might struggle with the complexity of photorealistic images. Other models that operate at a higher level of abstraction might generate more varied and creative results, but often lack the fine detail of pixel-by-pixel approaches. The selection of the right model is not just a technical decision; it's a creative one, heavily influenced by the specific requirements and desired aesthetics of the generated output.

Consider the differences between generating a cartoon and a highly realistic portrait. A simplistic model might suffice for a low-resolution cartoon, while a complex, state-of-the-art model would be necessary for a photorealistic image. This underscores the diversity within the field of generative models; each model is a tool with unique strengths and limitations, selected based on the specific task at hand. Matching the model to the task is a critical aspect of successful image generation.

Moreover, the quality of the generated images heavily relies on the quality and quantity of the training data. A model trained on blurry, low-resolution images will inevitably produce similar output. Conversely, a model trained on a massive, high-quality dataset will yield far more impressive and detailed results. This highlights the importance of data curation and preprocessing in generating high-quality synthetic images. Investing time and resources into preparing the training data is an investment in the quality of the final output.

Beyond simply generating images, generative models are capable of much more. They can be used to create variations of existing images, style transfer, even generating images from textual descriptions. These capabilities open up exciting new possibilities for creative expression and practical applications in fields like animation, game development, and even scientific visualization. The potential applications are vast and constantly expanding as the models themselves become more sophisticated.

The field of generative models is constantly evolving, with new techniques and architectures continuously being developed. New research leads to more advanced methods which improve the quality, efficiency, and capabilities of image generation. Therefore, remaining updated on the latest advancements is vital for professionals in this rapidly advancing field. Keeping abreast of these developments ensures you can leverage the most potent tools and techniques available.

Furthermore, understanding the underlying mathematics and algorithms is not strictly necessary for using many of these models, but understanding the basic principles of how they function will improve your ability to fine-tune and work with them effectively. Numerous software packages and libraries provide user-friendly interfaces, making complex algorithms accessible even to those without a strong mathematical background. However, having a foundational understanding provides a significant advantage in troubleshooting and optimizing the model's performance.

Finally, remember that even the most advanced generative models are ultimately tools. Their effectiveness depends on the user's skill and understanding. Experimentation and creativity are essential, and iterative refinement is key to achieving the desired results. The process of generating high-quality images is an interplay between technological capabilities and artistic vision. Embrace this interplay, and the results can be truly remarkable. The generative models are at your disposal; now it's up to you to create.

1.2 Early Image Generation Techniques

Early image generation, before the digital revolution's surge, relied on fundamentally different methods than those employed today. Think of the painstaking work of artists meticulously crafting images, pixel by pixel, often using tools far removed from the sophisticated software we have now. The process was slow, intensely manual, and the results, while sometimes breathtaking in their detail and artistry, were severely limited in terms of scalability and automation.

Consider the creation of simple patterns, often found in textiles or even early computer graphics. These were generated using basic algorithms, often involving repetitive procedures or simple mathematical formulas. A common example would involve generating a grid, then systematically filling in squares based on pre-defined rules or even random chance. The limitations were immediately apparent: complexity was hard to achieve, and any deviations from the initial rules demanded significant manual intervention.

The emergence of cellular automata, a revolutionary concept at the time, marked a significant step forward. These systems employed a grid of cells, each in a specific state, which evolved based on pre-determined local rules. Imagine it like a game of life, where the state of each cell (alive or dead) in the next generation depends on its current state and the state of its immediate neighbors. The beauty of cellular automata lay in their ability to generate surprisingly intricate patterns from simple rules, hinting at the potential of algorithmic approaches to generate complex visual structures. This ability to create complex systems from simple rules opened up entirely new creative possibilities.

Procedural generation, another pioneering technique, took a different approach. Instead of focusing on fixed rules like cellular automata, procedural generation focused on algorithms that could generate diverse outputs based on a set of parameters. Imagine creating a landscape using noise functions and elevation algorithms, where each execution produced a unique, yet believable, mountain range. This method, though computationally demanding for its time, promised greater control and variability in the generated images. The potential for creating realistic or fantastical environments captured the imagination of researchers and artists alike.

Early vector graphics also contributed to the landscape of image generation. Unlike pixel-based approaches, vector graphics represent images using mathematical descriptions of lines, curves, and shapes. Software like early CAD systems capitalized on this approach to allow for precise control and scalability. The result was sharp, clear imagery ideal for technical illustrations and diagrams where pixel-based techniques would have been too inefficient. The key benefit was the ability to easily zoom and scale images without any loss of quality, a significant advance over raster-based methods.

While crude by today's standards, these early techniques played a vital role. They laid the foundation for the sophisticated algorithms and deep learning models used in modern image generation. They proved that intricate and beautiful imagery could be generated not just by human hands, but through clever programming and algorithmic design, a significant intellectual and artistic leap forward in its own right. Their limitations are obvious from a modern perspective, but their ingenuity and farsightedness remain noteworthy accomplishments in the long and continuing history of image generation. They established the conceptual groundwork upon which the field is built today. It's a testament to the power of human creativity that such simple, foundational methods could lead to the astonishingly realistic and diverse images generated by today's advanced technology.

2 Pixel Art & Low-Resolution Methods

Creating pixel art involves manipulating individual pixels to form an image. Think of it like working with tiny colored squares on a grid. Unlike higher-resolution images, the limitations of low resolution force a unique artistic style; each pixel holds significant visual weight, demanding careful consideration of color placement and overall composition. This constraint fosters creativity, leading to distinct aesthetics rarely achieved in photorealistic work. The simplicity belies a surprising depth of expression.

Mastering basic algorithms is crucial. Simple random dot patterns, for instance, while seemingly elementary, can form the basis for more complex techniques. Experiment with varying the density and color distribution of these dots to explore different visual effects. You can generate intriguing textures and abstract forms surprisingly quickly, and these basic forms can build up into more complex patterns. Consider introducing simple rules to govern the dot placement, leading to the emergence of structure within the apparent randomness. These foundational exercises will lay the groundwork for a deeper understanding of more advanced methods.

Cellular automata, employing simple rules applied iteratively to a grid of cells, are a powerful tool. Each cell's state evolves based on the states of its neighbors. This seemingly straightforward concept generates complex, often unpredictable patterns. Consider the Game of Life, a classic example showcasing how simple rules can lead to emergent behavior. The elegance and complexity found in this seemingly simplistic system offer endless possibilities for visual exploration. By adjusting the initial conditions and rule sets, you can cultivate a vast range of visual outcomes. You might start with simple rules and then add more and more complexity slowly.

Procedural generation elevates algorithmic image creation. Instead of manually placing each pixel, you define rules and parameters that the computer uses to generate the image. This technique is particularly useful for creating large or repetitive patterns, such as textures or backgrounds. Think about generating a brick wall: define the size and color of each brick, and an algorithm can automatically arrange them to form a realistic-looking wall, scalable in size and complexity. This approach allows you to create vast and diverse environments with relative ease. This method isn't limited to static images; you can create dynamic, changing scenes by building in variables that change over time.

Combining these low-resolution methods adds another layer of creative power. Imagine layering a procedurally generated texture atop a cellular automaton pattern, then carefully adding detailed pixel art elements.

This approach creates rich visuals with layers of depth and complexity. Experiment with different combinations, iterating and refining your techniques. The beauty lies in the unexpected interactions between these various methods, generating unique and surprising results. By embracing these limitations, you'll find exciting possibilities.

Software specifically designed for pixel art offers tools tailored to this unique style. Many programs allow you to control individual pixels with precision, offering palettes and brushes that enhance the artistic workflow. The user interface will often cater to ease of use in this specific context. The digital tools can greatly assist the creative process. Exploring different software packages will reveal the variety of features and interfaces available to best suit your personal style and preferences. This helps the creative process and encourages artistic flow.

Beyond the technical aspects, pixel art benefits from artistic principles shared with all visual media. Consider color theory, composition, and contrast. A limited color palette can significantly influence the overall mood and feeling of the image, creating a unique aesthetic. Careful arrangement of elements within the image will help to create a visually appealing composition, directing the viewers' eyes to the key elements of the picture. Remember, effective contrast enhances the visual impact of your work. Understanding and applying these artistic considerations elevates your pixel art from merely technical exercises to compelling pieces of visual storytelling.

Finally, embrace experimentation. The field of pixel art is rich with possibilities. The constraints, paradoxically, encourage creativity and innovation. By repeatedly experimenting with different algorithms, adjusting procedural rules, and refining your pixel placement, you will quickly discover how to transform simple concepts into unique, expressive works of art. The journey of exploration and discovery is a key part of the creative process itself; enjoy the process.

2.1 Simple Algorithms

Random dot patterns form the bedrock of many simple algorithms for image generation. Imagine a canvas, a grid of pixels, each awaiting a fate determined by pure chance. A simple loop assigns each pixel a color—black or white, perhaps—based on a random number generator. The result? A seemingly chaotic arrangement, yet surprisingly, these random patterns can form the basis for more complex structures, acting as a foundational layer for subsequent processing steps such as noise reduction or the application of simple filters. The simplicity of this approach is its strength, allowing even the most novice programmer to immediately grasp and implement the concept. Experimentation with the density of dots, or the introduction of grayscale values, adds further complexity and creative possibilities.

Moving beyond random noise, we can introduce structured randomness. Consider a grid where each cell is assigned a value based on a simple mathematical function, perhaps a sine wave or a random walk. This approach adds a degree of coherence to the randomness, resulting in patterns that are less chaotic and more visually appealing. Think of shimmering textures, subtle gradients, or abstract landscapes—all attainable through judicious use of mathematical functions to modulate the random number generator. This level of control opens the door for creative exploration, allowing the artist to dictate the level of randomness and the overall aesthetic direction. Furthermore, combining various mathematical functions can lead to unexpected and novel visual outcomes. Such simple algorithms provide a surprisingly rich palette for visual exploration.

Let's delve into a slightly more sophisticated algorithm: the use of cellular automata. Cellular automata are mathematical models that use simple rules applied to a grid of cells to create complex patterns. The rules dictate how the state of a cell changes based on the state of its neighbors. A classic example is Conway's Game of Life, where simple rules governing birth, death, and survival of cells result in complex and evolving patterns. This can be directly translated into image generation by mapping cell states to colors. The initial configuration of the cells acts as the "seed" of the image, and the evolution of the system over time generates the final image. The beauty of this approach lies in its ability to create surprising complexity from simple rules, offering a powerful tool for artistic exploration and the generation of intricate textures. The parameters of the rules themselves become creative inputs, allowing for fine-grained control over the generated image's aesthetic qualities.

Another approachable technique utilizes a procedural approach. Procedural generation leverages algorithms to create content automatically. Imagine generating a landscape: instead of manually placing every tree and mountain, an algorithm could define the height of the terrain based on a mathematical function, then procedurally generate trees and other elements based on parameters such as terrain height, slope, and proximity to water sources. This process could even incorporate randomness to ensure diversity and avoid repetitiveness. The advantages of this method are numerous: infinite variability, efficient generation of vast amounts of content, and the ability to control parameters such as scale, detail, and complexity. Consider the immense possibilities for creating game worlds, generating textures for digital art, or even designing architectural structures – all with a few lines of code implementing a well-defined procedural generation algorithm.

Finally, let's explore the power of simple iterative techniques. Algorithms based on iterative processes build images step-by-step. An example could be an algorithm that starts with a single point or a simple shape, and repeatedly modifies it based on a set of rules. Think of branching structures like trees or snowflakes, where each branch is recursively generated by applying the same set of rules at smaller scales. The resulting fractal-like structures can be incredibly intricate and detailed, arising from surprisingly concise algorithms. This method is particularly well-suited for creating organic forms and textures that exhibit self-similarity at different scales, lending itself well to generating natural phenomena and abstract designs. The iterative refinement process, though computationally straightforward, provides the potential for an almost infinite complexity in the final outcome. The simplicity of the underlying algorithm allows for rapid prototyping and experimentation.

2.2 Cellular Automata

Cellular automata, at their core, are simple systems capable of generating surprisingly complex patterns. Imagine a grid, each cell capable of existing in a finite number of states, often just "on" or "off." The magic happens with rules: each cell's next state depends solely on its current state and the states of its neighbors. This seemingly simplistic setup can lead to intricate, evolving designs, perfect for generating textures, landscapes, and even abstract art in low-resolution image synthesis.

These rules, while simple, govern the entire evolution of the system. A common example is Conway's Game of Life, where a cell's fate hinges on the number of its live neighbors. Such rules, often expressed as mathematical functions, are the cornerstone of designing specific cellular automata. Consider the possibilities: patterns that spread, patterns that remain static, patterns that oscillate, all arising from the interplay of these fundamental rules. The key lies in carefully crafting the rules to achieve your desired visual outcome.

Experimentation is key. Start with a small grid and simple rules. Observe how different rule sets affect the pattern's evolution. You might begin with a single "live" cell in the center, allowing the rules to dictate the subsequent pattern. Then, try introducing more initial cells, scattered randomly across the grid. Each time, make a note of the evolution. The resulting visual patterns, while stemming from straightforward rules, will often showcase unexpected and intriguing complexities, revealing a unique visual language.

One practical application in image generation involves using cellular automata to create textures. By assigning different colors or shades to the different cell states, you can generate textures that mimic things like stone, wood, or even fabric. The choice of rules dramatically affects the appearance of the resulting texture. A rule that promotes rapid growth will create a bolder, more chaotic texture. Conversely, rules that limit growth can lead to subtler, more uniform patterns. This gives you precise control over the stylistic attributes of the textures you produce.

Programming cellular automata is surprisingly straightforward. Many programming languages offer suitable frameworks, making it a fantastic starting point for image generation projects. The process usually involves initializing the grid, defining the rules as a function, and then iteratively updating the grid based on these rules. This can involve loops and conditional statements, a core aspect of any computational process. You'll likely need to implement visualization techniques to display the grid and its evolution, perhaps using a suitable graphics library for enhanced visual outputs.

Beyond textures, consider utilizing cellular automata to generate simple landscapes. You might represent elevation with cell states, with higher states corresponding to higher elevations. Then, apply rules that simulate erosion or tectonic shifts. This could lead to the generation of realistic, albeit low-resolution, mountain ranges or valleys. By manipulating the initial conditions and rule sets, you can explore a wide variety of landscape characteristics. The ability to easily experiment is a key strength of this method.

The choice of rule set is crucial in shaping the final result. The creation of tailored rules often demands an iterative approach involving trial and error, testing various scenarios, and refining the process to obtain the desired pattern. This necessitates an understanding of how different rules propagate and interact, a crucial aspect for mastering cellular automata. Therefore, meticulous design and a deep understanding of the system's underlying dynamics are critical to achieving your generative goals.

Cellular automata's relatively simple nature makes them a powerful tool for teaching fundamental concepts in computational modelling and image generation. Their ability to produce complex outcomes from simple rules demonstrates the emergent properties of complex systems. It's a fantastic starting point for anyone interested in the creative potential of algorithms and computationally driven art. The ease of implementation, coupled with their unexpected visual output, makes them ideal for both novice and experienced programmers.

Remember, the beauty of cellular automata lies not just in their capability to generate complex patterns, but also in the process of experimentation and exploration. The iterative nature of designing and refining the rule sets provides ample room for creative expression and discovery. While seemingly simplistic at first glance, cellular automata present a rich and rewarding approach to image generation, revealing the generative power of simple systems. This direct connection between simple rules and complex visuals makes them an ideal tool for both learning and artistic exploration. Experiment, iterate, and let the emergent patterns guide you.

2.3 Procedural Generation

Procedural generation offers a compelling approach to creating diverse and repeatable image content algorithmically. Unlike hand-crafted assets, procedural methods leverage algorithms to generate variations of images, textures, or even entire game environments, saving artists significant time and effort. This is especially beneficial for creating large-scale content such as game worlds or environments where manual creation would be prohibitively time-consuming. Consider, for instance, the generation of a vast forest: hand-drawing every tree would be incredibly labor-intensive. However, a procedural algorithm could generate thousands of unique trees based on a set of parameters, resulting in a believable and varied forest landscape.

The core principle is defining a set of rules or parameters that govern the generation process. These rules can range from simple random number generators, determining the placement of elements, to sophisticated algorithms that model natural processes like branching patterns in trees or the formation of clouds. The flexibility lies in the complexity of the algorithm itself; a simple algorithm might produce simplistic results, whereas a more complex one offers greater control and nuance in the output, resulting in higher fidelity and more convincing results. The selection of parameters is crucial to the final aesthetic outcome.

This control extends beyond simple placement. Consider texture generation. Instead of meticulously painting each texture, algorithms can be designed to generate realistic wood grain or stone patterns by simulating natural processes at a microscopic level. This process allows for the creation of textures that vary naturally, eliminating repetition and improving visual interest. The textures generated can be seamlessly integrated into other elements of the procedural generation pipeline, creating a highly cohesive and visually appealing environment. This ability to generate not just varied instances but fundamentally new forms opens up exciting possibilities for creative expression.

For beginners, it's advisable to start with simpler procedural techniques. Learning to manipulate basic random number generators to place elements within a defined space is an excellent starting point. Experiment with different distributions and seed values to observe their influence on the generated output. Gradually, explore more complex methods such as L-systems, which are particularly effective for generating self-similar structures like plants and trees. Mastering these fundamental techniques is key to tackling more advanced aspects of procedural generation.

Implementing procedural generation often involves programming. While specialized software exists, understanding basic programming concepts, particularly loops and conditional statements, proves invaluable in designing and implementing procedural algorithms. Languages like Python offer many powerful libraries that significantly simplify the process. However, even a basic understanding of the underlying principles will empower you to adapt existing algorithms or create your own to suit your specific needs. Remember, building upon existing knowledge and iteratively refining your algorithms is crucial to effective procedural generation.

Moreover, consider the limitations of procedural generation. While it's incredibly powerful, it's not a silver bullet. The level of detail and control achievable is often tied to the complexity of the algorithm. Overly simplistic algorithms may produce repetitive or unconvincing results. Conversely, extremely complex algorithms can be difficult to implement, debug, and maintain. Finding the right balance between complexity and output quality is a key aspect of successful procedural content creation. This is achieved through iterative testing and refinement.

The beauty of procedural generation lies in its ability to seamlessly blend artistic control and algorithmic efficiency. By carefully selecting parameters and designing sophisticated algorithms, artists can leverage the power of code to create visually rich and diverse content that would be otherwise impossible to generate manually. The ability to generate unique content on demand provides the user with remarkable creative versatility. It's important to remember that procedural generation is not about replacing human creativity, but about augmenting it, empowering artists to explore new realms of expression and efficiency. This allows artists to focus on the creative aspects, leaving the repetitive tasks to the algorithms.

Ultimately, the success of procedural generation hinges on a nuanced understanding of algorithms, parameters, and the iterative refinement of both. Beginners must approach this field with a spirit of experimentation and a willingness to learn. Mastering the art of procedural generation unlocks a world of possibilities, empowering artists and designers to produce vast and compelling image content with remarkable efficiency and creative freedom, expanding their artistic vision and creative capacity exponentially. By combining technical skills with an artistic eye, one can create truly unique and visually stunning images.

3 Vector Graphics & Geometry

Scalable Vector Graphics (SVGs) offer a unique approach to image generation, unlike pixel-based methods. Instead of defining an image through a grid of colored pixels, SVGs utilize mathematical descriptions of shapes, paths, and text. This means the image resolution is independent of the file size; you can zoom in infinitely without losing quality. This makes SVGs ideal for logos, illustrations, and designs that need to be displayed at various scales without pixelation. Mastering SVGs involves understanding XML-based structure and the different elements it supports – lines, curves, rectangles, circles, and more. Furthermore, you can apply styles, gradients, and filters to enhance their visual appeal. The beauty lies in the precision and scalability. You can precisely control every aspect of each element, allowing for intricately detailed work.

Consider the power of manipulation: you can easily modify an SVG image after it's created, adjusting its dimensions, colors, or shapes without regeneration. This iterative design process is a significant advantage over pixel-based alternatives. This flexibility makes SVGs a highly efficient choice for projects requiring dynamic updates or edits. This editable nature reduces the need for repeated rendering or creation. Many vector graphics editors exist, offering intuitive interfaces and powerful tools for creating stunning visuals. Experiment with different software to find the one that best suits your workflow and preferences. Remember, practicing with different tools and techniques is key.

Fractals are another fascinating avenue within vector graphics. They are self-similar geometric patterns that repeat infinitely at different scales. Generating fractals involves iterative algorithms, often based on mathematical equations. Simple algorithms can produce surprisingly complex and visually striking results. The most famous example is the Mandelbrot set, generated from a seemingly simple formula, yet displaying an unfathomable complexity of shapes and colors. Explore the various fractal types—Julia sets, Sierpinski triangles, Cantor sets—each offering unique visual characteristics and mathematical underpinnings. Numerous software packages and libraries are readily available to aid in exploring and creating fractal imagery. You can start experimenting with readily available code snippets and gradually increase the complexity as your understanding improves.

L-systems, or Lindenmayer systems, offer yet another powerful technique for generating complex forms. These are string rewriting systems that use a set of rules to iteratively expand a starting string. This string is then interpreted to define a geometric shape. L-systems are particularly useful for modelling plants, trees, and other natural structures with branching patterns. The starting string defines the initial geometry, and the rewriting rules determine how this geometry develops over subsequent iterations. The elegance of L-systems lies in their ability to generate complex shapes with remarkably simple rules. By adjusting the rules and parameters, you can create a huge diversity of visual outcomes. Understanding the role of iteration in shaping the final form is a vital part of the process. Experiment with different rulesets and observe the remarkable diversity that emerges.

Combining SVGs with fractals and L-systems opens up a wealth of possibilities for creating truly unique and visually captivating images. Imagine using an L-system to generate a fractal tree structure, then importing this into an SVG editor to add detail, color, and text. This hybrid approach allows for fine-grained control over individual elements while maintaining the underlying generative power of the fractal or L-system. The possibilities are endless. Consider creating realistic-looking landscapes or fantastical alien flora; the limitations are purely your imagination and your knowledge of the tools at your disposal. You can explore several libraries and tools that combine these techniques, allowing you to streamline your workflow.

The integration of these methods transcends purely aesthetic pursuits. In game development, for instance, procedurally generated environments, characters, and items rely heavily on these techniques. Imagine creating a whole game world based on fractal landscapes and L-system vegetation, dynamically generated each time the game is played. The efficiency and scalability of vector graphics enable such ambitious undertakings. The ability to create complex, unique elements without sacrificing performance is a significant advantage in resource-intensive applications such as game development and interactive simulations. Moreover, the mathematical underpinnings of these methods make them ideal for applications beyond artistic expression, extending to scientific visualization and simulation.

3.1 SVG and Image Creation

Scalable Vector Graphics, or SVG, offer a unique approach to image creation, distinct from the pixel-based methods discussed previously. Unlike raster graphics, which are composed of a grid of pixels, SVG images are defined by vectors—mathematical descriptions of lines, curves, and shapes. This vector-based nature grants SVGs several key advantages: they are resolution-independent, meaning they can be scaled to any size without losing quality; they are highly editable, allowing for precise adjustments to individual elements; and they generally produce smaller file sizes compared to equivalent raster images, making them ideal for web applications and other contexts where efficiency is paramount. Understanding this fundamental difference is crucial for harnessing their full potential.

The core of SVG manipulation involves using specialized software or coding languages like XML. This contrasts sharply with the more intuitive, visual interfaces often associated with raster graphics editors. You'll find yourself working directly with the mathematical representation of your images, specifying coordinates, curves, and fills for each element. This seemingly abstract approach yields exceptional control and precision, enabling the creation of complex illustrations and animations with remarkable fidelity. However, this level of precision often requires a steeper learning curve than raster graphics. Mastering the nuances of SVG requires dedicated study and practice.

Think of creating an SVG illustration as building with digital LEGO bricks. Each shape, line, and curve is a separate building block, meticulously placed and manipulated. You can easily alter the size, color, or position of individual components without affecting others, allowing for precise refinement and iterative design. For instance, you could create a complex logo by assembling individual vector shapes—circles, rectangles, and paths—then grouping them to form a cohesive whole. This approach is particularly valuable when creating scalable logos or illustrations that need to look perfect across various resolutions, from tiny website icons to large billboards.

But SVG isn't limited to simple shapes. Advanced techniques allow for complex gradients, patterns, and effects, opening up a world of creative possibilities. Mastering these techniques enables the crafting of detailed illustrations, intricate designs, and even dynamic animations. For example, you could create a responsive, animated web banner where elements smoothly transition and interact based on user input or predefined timing sequences. This type of intricate animation is only possible through the precision and flexibility of SVGs.

While the initial learning curve might seem daunting, abundant resources exist to aid beginners. Online tutorials, interactive courses, and extensive documentation are readily available. Starting with simple shapes and gradually incorporating more complex elements is a proven strategy. Begin by experimenting with basic geometric forms. Gradually introduce more advanced features such as gradients, filters, and transformations as you gain familiarity with the fundamental concepts. Remember, practice is key to mastering any new skill. Consistent practice strengthens both your understanding of the underlying principles and your proficiency in creating stunning SVG-based visuals.

Moreover, the ability to create SVGs is a highly sought-after skill in the professional design field. Whether it's web design, graphic design, or animation, the capacity to produce scalable and highly editable vector-based assets is immensely valuable. Acquiring SVG skills empowers you with a significant advantage in today's digital marketplace, opening doors to diverse career paths and creative endeavors. Therefore, the investment of time and effort in mastering SVG creation will likely yield substantial returns in both professional and personal contexts. It provides a powerful and versatile toolset for creating visually stunning and technically efficient images. The adaptability and scalability of SVG make it a crucial technology for creating contemporary digital graphics. The collaborative nature of SVG development further enhances its appeal. Many open-source tools and libraries are available, facilitating teamwork and knowledge sharing amongst developers. This collaborative environment accelerates the development process and fosters innovation within the SVG community. Therefore, the continuous development and refinement of SVG technology ensure its longevity and continued relevance in the broader landscape of digital image creation. Embrace this dynamic and versatile tool, and unlock a whole new dimension of creative potential.

3.2 Fractals and L-Systems

Fractals, with their self-similar patterns repeating at different scales, offer a unique approach to image generation. Imagine a coastline; viewed from afar, it's a simple line. Zoom in, and intricate details, bays, and inlets emerge, mirroring the larger shape. This self-similarity is the core principle behind fractal generation, allowing for the creation of incredibly complex images from simple recursive algorithms. We'll explore how this principle translates into practical image generation, moving beyond simple theoretical concepts to hands-on application.

The Mandelbrot set, a classic example, is generated by iterating a simple equation. The result? A stunningly complex image teeming with intricate detail, a visual testament to the power of iterative processes. Understanding the underlying mathematical principles—complex numbers and iteration—is key to unlocking the generative potential of fractals. Mastering these fundamentals enables you to customize the algorithm, generating unique and captivating images tailored to your specific needs. This isn't merely about replicating existing fractals; it's about learning to design your own.

L-systems, or Lindenmayer systems, provide another compelling path toward generative image creation. These algorithmic systems, inspired by the growth patterns observed in plants, use string rewriting rules to generate complex branching structures. Start with an initial string, like "A," and apply a set of rules to iteratively rewrite it. For instance, "A" might become "AB," "B" might become "A." Repeating this process creates increasingly intricate strings, which can then be interpreted geometrically to create visual representations. This process naturally lends itself to creating organic shapes, ideal for vegetation, trees, or even abstract art.

The beauty of L-systems lies in their flexibility. By tweaking the rules, you can control the branching angle, the length of each branch segment, and other parameters, fine-tuning the final image. You can experiment with different rule sets to explore new visual territories. Imagine the possibilities: creating stylized trees for game environments, generating unique patterns for fabric designs, or even designing complex architectural structures. The potential applications are vast and largely unexplored. It's a fertile ground for artistic exploration and experimentation.

Combining fractals and L-systems opens up even more creative avenues. For instance, you could use an L-system to generate the branching structure of a tree, then apply fractal algorithms to create detailed textures on the leaves or bark. This hybrid approach allows for the generation of highly realistic and complex organic forms, extending the capabilities of each technique independently. Such integration showcases the synergistic power of combining different generative techniques for producing novel results. It transcends the limitations of using a single method, opening up a richer design space.

Programming these algorithms is more accessible than you might think. Many programming languages offer libraries and frameworks designed specifically for fractal and L-system generation. Understanding basic programming concepts, loops, recursion, and string manipulation are sufficient to start generating your images. Numerous online resources and tutorials are available to guide you, offering practical examples and code snippets that will help you quickly get your hands dirty and start producing amazing visualizations. There's a rich community built around procedural generation, providing invaluable support and inspiration.

Don't be intimidated by the mathematics. While an understanding of the underlying concepts is helpful, you don't need to be a mathematician to generate impressive results. Start with simple examples, experiment with different parameters, and gradually increase the complexity as you become more comfortable. The key is to iterate, experiment, and observe. Each change in the code translates to a visible modification in the generated image—instant feedback that fuels the creative process. The process of creation is an experiment in itself, a journey of visual discovery.

Remember, this chapter serves as a starting point. The world of fractals and L-systems is vast and continues to evolve, fueled by ongoing research and innovation. This is not an exhaustive guide, but rather a pathway towards engaging with these powerful tools. Explore, experiment, and most importantly, have fun! The generative potential of these techniques is immense, waiting for your exploration and innovative approach. The beauty of this chapter is its ability to empower you to create unique and stunning visual imagery, an image that represents your own creative vision, generated not just by algorithms, but by your creative engagement with them. Use your newly acquired knowledge to discover something novel and captivating.

4 Deep Learning for Image Generation

Generative Adversarial Networks, or GANs, are a cornerstone of modern image generation. Imagine two neural networks locked in a constant creative battle: a generator, striving to create convincing fake images, and a discriminator, meticulously judging their authenticity. This adversarial process pushes both networks to improve, resulting in increasingly realistic outputs. The generator learns to fool the discriminator by crafting more detailed and nuanced images, while the discriminator sharpens its ability to identify forgeries. This dynamic, competitive training is what fuels the power of GANs. Understanding this core mechanism is paramount to leveraging their capabilities effectively. Successfully training a GAN requires careful consideration of various hyperparameters, network architectures, and data preprocessing techniques. The choice of architecture, for instance, can significantly impact the quality and stability of the generated images. Deep Convolutional GANs (DCGANs) utilize convolutional layers to capture spatial hierarchies in images, leading to more coherent results. Similarly, the selection of an appropriate loss function—a measure of the network's performance—influences the training process and the ultimate quality of the generated images.

Variational Autoencoders, or VAEs, offer a different approach. Instead of the adversarial training of GANs, VAEs employ a probabilistic framework. They learn a compressed representation of the input data, capturing the underlying latent features. This latent space can then be sampled to generate new images. Think of it as learning the essence of an image and then using that essence to create variations. VAEs excel in tasks where controllable generation is crucial. By manipulating the latent variables, one can subtly alter the characteristics of the generated image, offering a level of control not always achievable with GANs. However, VAEs often struggle to generate high-resolution images with fine details, a limitation that contrasts sharply with the capabilities of some advanced GAN architectures. The quality of generated images from VAEs is highly sensitive to the complexity of the model and the size of the training dataset.

Diffusion models have recently emerged as a powerful alternative, demonstrating exceptional results in generating high-quality images. These models work by gradually adding noise to an image until it becomes pure noise, and then learning to reverse this process. Imagine starting with a perfectly clear photo and slowly blurring it until it's indistinguishable from random pixels. A diffusion model learns how to reverse this process, starting from noise and gradually refining it into a coherent image. The advantage here lies in their ability to produce very detailed and sharp images. Furthermore, diffusion models tend to be more stable during training than GANs, requiring less meticulous hyper-parameter tuning. A major drawback is the computational cost associated with training these models; they require significant computational resources and time.

Autoregressive models generate images pixel by pixel, predicting the value of each pixel based on the previously generated pixels. This sequential approach allows for precise control over the generated image. Imagine constructing a house brick by brick—each brick placement depends on the previous ones. Autoregressive models work in a similar fashion, ensuring a coherent and logically structured output. Their strength lies in the fidelity of the generated images and their ability to model complex relationships within the data. However, the sequential nature of generation makes these models computationally expensive, particularly for high-resolution images. The training process can also be quite intricate, often requiring careful consideration of various architectural choices.

Style transfer techniques allow the application of the style of one image to the content of another. Imagine transforming a photograph of a landscape into a Van Gogh painting, preserving the original landscape's content but expressing it in the unique artistic style of Van Gogh. This is achievable through deep learning models that separate the content and style information in images and then recombine them to create novel outputs. These techniques are invaluable for artistic applications, offering a powerful tool for creative exploration and experimentation. Different architectures and methods exist for style transfer, each with its own strengths and limitations. Choosing the right approach often depends on the specific application and the desired level of control over the resulting image. A deep understanding of the underlying mechanisms is necessary for effective implementation and optimization.

4.1 Generative Adversarial Networks (GANs)

Generative Adversarial Networks (GANs) represent a significant leap forward in image generation. Unlike earlier methods that relied on explicit rules or pre-defined patterns, GANs leverage the power of competing neural networks to create remarkably realistic and diverse images. At their core, two networks—a generator and a discriminator—engage in a dynamic adversarial process. The generator attempts to produce convincing synthetic images, while the discriminator strives to distinguish these fakes from real images from a training dataset. This ongoing competition pushes both networks to improve, resulting in increasingly realistic outputs from the generator. The generator learns to better mimic real images, and the discriminator becomes more adept at spotting imperfections. This iterative process, akin to an artistic arms race, is what allows GANs to generate such compelling results.

Imagine a counterfeiter striving to create perfect banknotes, constantly challenged by an expert detective dedicated to uncovering forgeries. The counterfeiter refines his techniques, learning to replicate intricate details and security features, while the detective hones his skills in identifying subtle inconsistencies. This analogy, while simplified, captures the essence of the adversarial training process within a GAN. The continuous feedback loop between the generator and the discriminator drives improvement, leading to increasingly sophisticated outputs. The quality of the generated images hinges crucially on the training data—a large, diverse dataset is essential for generating realistic and varied results. Insufficient or biased data will invariably result in limited or skewed outputs, reflecting the limitations of the training set.

Training GANs can be notoriously challenging. Achieving stability and avoiding mode collapse, a phenomenon where the generator produces only a limited set of similar images, requires careful attention to hyperparameters and network architecture. Different GAN architectures, such as Deep Convolutional GANs (DCGANs) and Wasserstein GANs (WGANs), offer various advantages and disadvantages, influencing the stability, efficiency, and quality of the generated images. Choosing the right architecture and carefully tuning the training parameters often involves extensive experimentation and a deep understanding of the underlying principles. This requires a significant level of expertise and computational resources, and this can greatly influence the success or failure of a GAN training project.

Beyond the core architecture, several advanced techniques contribute to improving GAN performance. Techniques such as spectral normalization, gradient penalty, and the use of specific loss functions can significantly enhance stability and generate high-quality images. These methods aim to address common issues like vanishing gradients and mode collapse, thereby improving the overall quality and diversity of the generated images. The optimization process itself is crucial; researchers continuously explore new optimization strategies to improve convergence and enhance image quality and prevent common training problems. The choice of optimization algorithms like Adam or RMSprop can also play a considerable role in GAN training.

The applications of GANs are remarkably broad and continue to expand. Beyond simple image generation, GANs have been successfully applied to diverse fields, including image-to-image translation, super-resolution, and style transfer. In image-to-image translation, GANs can transform images from one domain to another, such as converting sketches to photorealistic images or changing the season in a landscape photograph. In super-resolution, GANs can enhance the resolution of low-resolution images, producing sharp, high-quality outputs. Finally, style transfer allows applying the style of one image onto another, such as transferring the artistic style of a famous painter to a photograph. This versatility underscores the power and transformative potential of GANs across numerous domains.

The impact of GANs extends beyond basic image manipulation. Researchers are actively exploring the use of GANs in creating realistic synthetic data for training other machine learning models, particularly in fields with limited real-world data. By generating large amounts of synthetic data that mimics the characteristics of real data, GANs help overcome the limitations of small datasets. Moreover, GANs are playing an increasingly important role in areas like medical imaging, where synthetic data can be used to augment real datasets and to train models for diagnosing diseases with greater accuracy. The development of new architectures and training techniques continues to improve the capabilities of GANs and their applicability to an ever-widening range of problems. The ongoing development of GANs is likely to further accelerate their applications in various fields in the years to come.

4.2 Variational Autoencoders (VAEs)

Variational Autoencoders (VAEs) offer a compelling approach to image generation, differing significantly from GANs in their underlying methodology. Instead of pitting two networks against each other, VAEs learn a compressed representation of the input data, encoding it into a lower-dimensional latent space. This latent space acts as a blueprint for generating new images; by sampling from this space, we can create novel outputs that share characteristics with the training data. The process involves two main components: an encoder, which maps input images to the latent space, and a decoder, which reconstructs images from latent representations. Crucially, VAEs incorporate a probabilistic framework, acknowledging the inherent uncertainty in representing complex data like images. This probabilistic element, expressed through the use of probability distributions, is what sets them apart and allows for a more nuanced understanding of the underlying data structure.

Unlike GANs, which can be notoriously difficult to train, VAEs generally exhibit greater stability during training. This is because they learn to reconstruct input data, allowing for more straightforward monitoring of progress and easier detection of potential issues. Their optimization focuses on minimizing the reconstruction error, which is a more well-defined objective than the adversarial game present in GAN training. This simplified optimization process usually leads to faster convergence and better reproducibility of results. However, the quality of images generated by VAEs can sometimes be slightly lower than what's achievable with GANs, especially in terms of high-frequency details and sharpness. The trade-off, therefore, is between easier training and potentially slightly lower image quality.

One crucial aspect of implementing VAEs is choosing the right architecture for both the encoder and the decoder. These are often convolutional neural networks (CNNs), particularly effective at handling the spatial relationships within images. The encoder processes the image, compressing it into a lower-dimensional vector within the latent space. This vector is a compact representation of the image's key features. The decoder, on the other hand, takes this compressed representation and expands it back into an image of the original dimensions. The effectiveness of a VAE hinges upon the quality of this compression and reconstruction process. Poorly chosen architectures can lead to blurry or distorted images, even if the training process runs smoothly. Experimentation with different CNN architectures, including variations in layer depth, filter sizes, and activation functions, is often necessary to achieve optimal results.

The probabilistic nature of VAEs is embodied in the use of probability distributions within the latent space. This is generally modeled as a Gaussian distribution, where each dimension of the latent vector is associated with a mean and a variance. Sampling from this distribution introduces randomness into the image generation process, allowing for greater variation in the generated outputs. This is a key strength of VAEs, particularly useful in scenarios where generating diverse images is desirable, for example, generating images of faces with varying expressions or features. The mathematical underpinnings are rooted in probability theory, requiring a fundamental understanding of probability distributions, likelihood functions, and optimization techniques like variational inference. Grasping these concepts is crucial for building an intuitive understanding of VAEs and their capabilities.

Beyond image generation, VAEs find practical applications in various image-related tasks. They can be utilized for image denoising, inpainting, and even anomaly detection. In image denoising, for example, a corrupted image is fed into the VAE, which then aims to reconstruct a cleaner version, effectively removing noise. Similar principles apply to image inpainting, where missing parts of an image are filled in based on the latent representation learned by the VAE. For anomaly detection, a VAE is trained on normal images, and images that generate high reconstruction errors are flagged as potential anomalies. This versatility makes VAEs a highly valuable tool in various image processing domains beyond simply generating new images. The flexibility of VAEs allows for adaptation and customization to specific problems, increasing their practical applicability across many fields. The ability to tailor the architecture and training process to match the specific demands of the given task is what separates a successful VAE implementation from an unsuccessful one.

In conclusion, Variational Autoencoders provide a powerful and relatively stable alternative to GANs for image generation. Understanding the fundamental concepts of latent spaces, probabilistic modeling, and the interplay between encoders and decoders is crucial for successful implementation. Through careful selection of architecture and optimization techniques, you can harness the generative power of VAEs to create novel and diverse images, as well as apply them to various other image-related tasks, thereby expanding their usefulness and value in numerous applications. The combination of its relative ease of training and diverse applications underscores its importance in the field of image generation and beyond. By mastering this technique, you gain another powerful tool in your arsenal for the creative manipulation of images.

4.3 Diffusion Models

Diffusion models represent a compelling approach to image generation, offering a unique perspective compared to GANs or VAEs. Unlike adversarial training, diffusion models operate by iteratively adding noise to an image until it becomes pure Gaussian noise, and then reversing this process to generate new images. This forward diffusion process is deterministic, making it straightforward to implement. The reverse diffusion process, however, is learned, requiring a neural network to progressively remove the noise and reconstruct the image. This learning process hinges on minimizing the discrepancy between the generated image and the training data.

The core concept behind the success of diffusion models lies in their ability to capture intricate details and subtle variations within the training data. Unlike GANs, which often suffer from training instability and mode collapse, diffusion models tend to be more stable and generate a wider variety of images. They achieve this through the gradual refinement process inherent in the reverse diffusion. Each step involves a small denoising operation, allowing the model to learn and correct for errors incrementally. This iterative nature lends itself to generating high-quality images with sharp details and realistic textures, surpassing GANs in certain image fidelity metrics.

This iterative refinement, however, comes at a computational cost. Generating a single image using diffusion models typically requires many more steps compared to other generative models. This computational burden is a significant hurdle, particularly when working with high-resolution images. Researchers continue to actively explore optimizations and efficient architectures to mitigate this limitation, but it remains a critical factor to consider when choosing a generative model for a specific application. The trade-off between image quality and computational efficiency needs careful evaluation.

One key advantage of diffusion models lies in their flexibility. The underlying framework can be adapted to various data modalities beyond images, such as audio and video. This adaptability is driven by the fundamental principle of the diffusion process, which transcends specific data formats. Furthermore, diffusion models can be conditioned on various inputs, such as text prompts, allowing for fine-grained control over the generated content. This opens up exciting possibilities for creative applications and task-specific image generation, moving beyond simple random generation.

Implementing a diffusion model involves choosing an appropriate neural network architecture for the denoising process. Common choices include U-Nets, known for their ability to capture both global and local features within the image. The training process itself requires a substantial amount of data, mirroring the demands of other deep learning models. Careful selection of hyperparameters, such as the number of diffusion steps and the learning rate, is crucial for optimal performance. Experimentation and fine-tuning are essential steps in achieving satisfactory results, requiring patience and iterative adjustments.

Beyond the technical aspects, understanding the inherent biases present within the training data is paramount. Diffusion models, like other generative models, can inherit and amplify biases present in the datasets they are trained on. This can lead to the generation of images that reflect societal biases, potentially reinforcing harmful stereotypes or perpetuating unfair representations. Careful consideration of dataset curation and the development of bias mitigation techniques are crucial to ensuring responsible and ethical use of diffusion models.

Successfully leveraging diffusion models requires understanding not just the technical details but also their limitations and potential pitfalls. While they excel in generating high-quality images, their computational demands can be a significant constraint. Moreover, addressing inherent biases within training data is an ongoing challenge that needs to be tackled proactively. However, the continuous research and development within this field suggest that diffusion models will play an increasingly important role in image generation, offering a powerful and versatile tool for various creative and practical applications. The ability to generate high-fidelity images and adapt the model to different data formats makes diffusion models a valuable asset in the growing landscape of image synthesis. The potential for further advancements and optimizations makes this approach exceptionally promising for the future of image generation. Further research is needed to fully address the computational cost and to explore even more creative applications of this promising technology. The future of image generation, it seems, is brimming with possibility.

4.4 Autoregressive Models

Autoregressive models offer a unique approach to image generation, differing significantly from GANs or VAEs. Instead of directly generating an entire image at once, they predict one pixel at a time, sequentially building the image from left to right and top to bottom, conditioned on the previously generated pixels. Imagine painting a picture, stroke by stroke, each brushstroke guided by what's already on the canvas. This sequential nature makes these models surprisingly intuitive, mirroring how humans might approach creative tasks. They excel at capturing intricate details and complex dependencies between pixels, leading to highly realistic outputs in many applications.

The core principle behind autoregressive models lies in their probabilistic nature. Each pixel is predicted as a probability distribution over possible color values. This allows the model to capture uncertainty and generate diverse outputs, even given the same initial conditions. This is a crucial distinction from deterministic methods; the inherent stochasticity of autoregressive generation allows for greater creativity and avoids repetitive outputs, creating a richer and more diverse image collection. The models don't just copy training data; they learn the underlying statistical distribution to effectively generate novel images that aren't exact copies from their training set.

One significant advantage is their ability to handle high-resolution images more effectively than some other generative methods. This stems from the sequential generation process; high-resolution images are constructed pixel by pixel, mitigating the computational challenges associated with generating large images simultaneously. However, the sequential generation process naturally introduces computational costs. Generating a single high-resolution image can be significantly slower than with methods like GANs, which generate the whole image at once, although recent optimizations continue to improve efficiency and speed. The trade-off between computational cost and image quality is often a central consideration when selecting a generative model.

A common architecture used for autoregressive image generation is the PixelCNN. This model employs a convolutional neural network, focusing on the local dependencies of neighboring pixels. By applying convolutional filters, the model captures spatial correlations effectively, ensuring the generated image maintains spatial coherence and avoids creating unrealistic artifacts, like inconsistencies or blurring, that can plague other models. The convolutional layers progressively extract features from the image, enabling the model to learn complex relationships, which directly translate into a more refined and detailed output image.

Training autoregressive models typically requires substantial computational resources and a large, high-quality dataset. The sequential nature of the generation process implies that the training process is inherently iterative and computationally intensive, requiring significant processing power and time. The model needs to learn the complex dependencies between pixels, demanding a large dataset to cover a wide range of possible image characteristics. While the quality of the generated images is generally high, the demands of training may restrict the accessibility of these models to individuals or groups with limited computational capabilities.

Despite the computational challenges, the advantages of autoregressive models remain compelling. Their ability to generate high-quality, realistic, and diverse images with precise details makes them valuable in various applications, including image editing, image enhancement, and style transfer. They offer a nuanced approach to image synthesis, allowing the model to learn and capture very fine details and correlations between neighboring pixels. Unlike some methods that focus more on global patterns, autoregressive models excel at capturing fine detail in local areas and constructing high-resolution images with exceptional precision. Their strengths offer a powerful alternative, particularly in areas requiring exceptional image quality.

Furthermore, the sequential generation process provides a certain level of control. One can in principle modify the generation process at various stages by introducing conditional information, or modifying the probability distributions during the generation of intermediate parts of the image. This allows for a degree of fine-grained control not always available with other generative models. This degree of control might become exceptionally relevant in specific applications requiring high levels of customization and user intervention. It permits tailoring the generation process to specific needs, offering significant advantages in creative applications or specific image manipulation tasks. However, achieving this level of control requires a deep understanding of the model's inner workings.

In conclusion, while computationally intensive and demanding in terms of data requirements, autoregressive models offer a unique and powerful approach to image generation. Their sequential, probabilistic nature enables the creation of high-quality images with exceptional detail and diversity. This makes them a valuable addition to the generative model toolkit, particularly when precise detail and control are paramount. While challenges remain in terms of efficiency and scalability, ongoing research continues to refine these models, expanding their applications and accessibility. The intricate details and capabilities of autoregressive models position them as a crucial component of the generative image synthesis landscape, offering a distinct path towards creative and realistic image synthesis.

4.5 Style Transfer Techniques

Style transfer, at its core, involves transferring the artistic style of one image onto the content of another. Imagine transforming a photograph of a mundane cityscape into a vibrant Van Gogh-esque painting, or applying the texture and color palette of a Monet water lily to a portrait. This isn't mere image manipulation; it's about algorithmically understanding and recreating artistic expression. The process involves disentangling the content and style representations within an image, a complex task that has seen significant advancements with the rise of deep learning.

Early methods relied on hand-crafted features and statistical approaches, focusing on low-level image characteristics like color histograms and texture analysis. While effective in simple cases, these lacked the nuance to handle complex stylistic variations. They struggled with the subtle interplay of brushstrokes, color harmonies, and composition that defines artistic style. Their limitations became glaringly obvious when applied to intricate styles.

The advent of deep learning revolutionized style transfer. Convolutional Neural Networks (CNNs), with their ability to learn hierarchical representations of images, provided a powerful tool for tackling this challenge. A seminal approach, known as Neural Style Transfer, utilized a pre-trained CNN – often a model trained for image classification like VGG-19 – to extract features from both the content and style images. The network learns to represent the content image's structural information and the style image's textural and color features separately.

A key concept here is the loss function. The algorithm minimizes the difference between the generated image's features and the target content and style features. This often involves a weighted combination of content loss and style loss, allowing for a balance between preserving the original image's content and applying the desired style. This optimization process iteratively refines the generated image until it aligns with both the content and style representations learned from the input images.

However, this initial approach wasn't without its limitations. Computationally, it was rather expensive and slow. Moreover, the generated images often exhibited artifacts or inconsistencies, particularly in regions with complex detail. Subsequent advancements focused on addressing these issues. More efficient architectures were developed, reducing computational complexity. Improvements to loss functions helped minimize artifacts and improve the overall quality of the transferred style.

Recent developments have explored various architectures and approaches. Some techniques leverage generative adversarial networks (GANs) to enhance the realism and coherence of style transfer results. GANs, by pitting a generator network against a discriminator network, help refine the generated image, ensuring that the transferred style is both visually appealing and consistent with the original content. They offer improved control over the style transfer process and often lead to higher quality outputs.

Furthermore, advancements have extended style transfer beyond still images. Researchers are now exploring the application of style transfer to videos, allowing for the transfer of artistic styles to moving images. This involves tackling the temporal consistency of style transfer, ensuring a smooth and coherent application of style across consecutive frames. This opens up new possibilities in film editing and animation. The algorithms must not only transfer styles but do so in a manner that matches the motion and dynamism of the video content.

Beyond artistic applications, style transfer finds its place in various fields. In medical imaging, style transfer can enhance the visibility of specific features in medical scans. In satellite imagery, transferring styles can improve the visual representation of geographic data. The possibilities are ever-expanding. Style transfer remains a vibrant area of research, constantly evolving with new architectures, refined loss functions, and increased efficiency. It offers a creative and powerful tool for artists, researchers, and designers alike. As computational power continues to grow and algorithms become more sophisticated, the applications of style transfer are bound only by the imagination. The potential to explore entirely new aesthetic realms using style transfer is immense.

5 Advanced GAN Architectures

StyleGAN, a groundbreaking architecture, revolutionized image generation with its innovative approach. Instead of directly generating images, StyleGAN employs a mapping network that transforms a latent vector into a series of style vectors. These styles, at different resolutions, guide the synthesis process, providing unprecedented control over various image features. This allows for nuanced manipulation of attributes like pose, lighting, and even fine details, resulting in incredibly realistic and high-fidelity outputs. The disentanglement of features, a key advantage, enables independent control, facilitating targeted modifications without unwanted side effects. Imagine effortlessly altering the hairstyle of a generated person without affecting their facial expression – that's the power of StyleGAN's meticulously crafted architecture. Further advancements, such as StyleGAN2 and StyleGAN2-ADA, built upon this foundation, addressing limitations and enhancing performance even further, refining the image quality and stability, minimizing artifacts, and improving training efficiency. This architectural refinement represents a significant leap forward in GAN capabilities. These iterative improvements underscore the dynamic nature of research in this field, consistently pushing the boundaries of realistic image synthesis.

The training of GANs, however, remains notoriously challenging. The adversarial nature of the generator and discriminator often leads to instability, manifesting as mode collapse, where the generator produces limited variations, or training failure, halting the entire process. Effective training necessitates careful consideration of hyperparameters, the selection of appropriate loss functions, and the implementation of robust regularization techniques. Regularization methods, such as weight decay and spectral normalization, help prevent overfitting and improve generalization, ensuring the model learns to generate diverse and high-quality images. Addressing this instability requires sophisticated strategies. One approach involves employing different loss functions, balancing realism and diversity in the generated images; another focuses on architectural enhancements, improving information flow and gradient stability. Successful GAN training frequently demands experimentation and a deep understanding of the underlying mathematical principles. The need for sophisticated techniques underlines the complexities involved in mastering the training process and consistently achieving desired results. Mastering this nuanced process requires both theoretical understanding and practical expertise.

Beyond StyleGAN, alternative architectures have emerged, each addressing specific shortcomings or exploring novel approaches. Progressive Growing of GANs (PGGAN), for example, starts with low-resolution images and progressively increases resolution, simplifying training and improving image quality. This method avoids the challenges associated with training high-resolution generators directly, incrementally building complexity and enhancing the realism of the generated images. Other variations incorporate conditional generation, allowing control over the generated image's content. This is achieved by incorporating class labels or textual descriptions into the input, guiding the generator to produce images aligned with the provided information. This allows for the creation of tailored images based on specified criteria, expanding the range of applications. Each innovative architecture contributes to the ever-evolving landscape of GAN technology, constantly pushing the boundaries of what's achievable in the field of image generation. We see a clear progression in the sophistication and effectiveness of each architectural approach.

Improving the stability of GANs is an ongoing area of active research. Strategies include employing different optimizers, such as AdamW, which often exhibits better convergence and stability compared to traditional Adam optimizers. Careful selection of hyperparameters such as learning rate and batch size is crucial in achieving stable training. Moreover, techniques like gradient penalty and Wasserstein distance have been shown to mitigate the issue of vanishing gradients, a major contributor to training instability. These approaches effectively regularize the training process, preventing mode collapse and promoting the generation of diverse and high-quality samples. Regularization, crucial in preventing overfitting and promoting generalization, is essential in maintaining the stability and effectiveness of the model. Each of these refinements demonstrates a concerted effort to overcome the challenges inherent in training generative adversarial networks. The pursuit of stability remains a central theme in the continued development of advanced GAN architectures.

The future of advanced GAN architectures is brimming with potential. Research is actively exploring novel architectures, novel loss functions, and new training techniques. The integration of GANs with other deep learning models promises to unlock even more creative and powerful image generation capabilities. For instance, the combination of GANs with diffusion models is already showing very promising results. Combining the strengths of both models could lead to significant improvements in image quality, control, and diversity, pushing the boundaries of what's currently feasible. The development of more efficient training methods, particularly those that reduce computational costs and energy consumption, is crucial for broadening the accessibility and applicability of GANs. The rapid advancements and continuous exploration in this vibrant field promise even more compelling breakthroughs in the years to come, further solidifying their role in the expanding realm of image generation.

5.1 StyleGAN & Beyond

StyleGAN, introduced by NVIDIA, represented a significant leap forward in GAN architecture. Its innovative style-based generator allowed for unprecedented control over the image generation process, enabling manipulation of various aspects like pose, lighting, and overall style independently. This disentanglement of image features proved highly effective in producing high-fidelity images with remarkable detail and consistency. Unlike previous GANs prone to artifacts and training instability, StyleGAN demonstrated superior image quality and improved stability, setting a new benchmark in the field. This breakthrough paved the way for subsequent advancements.

However, StyleGAN wasn't without its limitations. While producing high-quality images, it often exhibited subtle artifacts, particularly in the form of recurring patterns or textures. Researchers identified these as "artifacts" stemming from the generator's learned internal representations. These weren't visually jarring, but they hinted at room for improvement in the underlying architecture and training processes. Identifying and addressing these subtle flaws became a key focus for the following generations of GANs.

Building upon StyleGAN's success, StyleGAN2 addressed many of its predecessors' shortcomings. The most significant improvement lay in its architecture modification designed to mitigate the aforementioned artifacts. By refining the network's internal operations and integrating novel techniques, StyleGAN2 demonstrably reduced these artifacts, further enhancing the realism and fidelity of the generated images. This refinement showcased the iterative nature of GAN development, where identifying and rectifying limitations drives continuous progress.

Subsequent iterations, including StyleGAN2-ADA and StyleGAN3, continued this trend of refinement. StyleGAN2-ADA introduced an adaptive discriminator augmentation technique, improving training stability and generalization capabilities. This adaptive approach dynamically adjusts the training process based on the discriminator's performance, leading to more robust and reliable results. The inclusion of adaptive techniques highlighted the growing understanding of the intricate dynamics within GAN training. StyleGAN3 then focused on improving image quality and computational efficiency, making it more accessible for practical applications.

The development of StyleGAN and its successors didn't end with simply improving image quality; it significantly influenced the broader field of generative models. The style-based architecture, once a unique feature of StyleGAN, has since become a common motif in other generative models. Its ability to disentangle latent spaces—essentially separating different features of an image— has proved highly beneficial in other architectures. This widespread adoption testifies to StyleGAN's lasting impact on the landscape of image generation techniques.

Beyond specific architectural improvements, the success of StyleGAN also underscored the importance of carefully curated datasets. The quality of the training data directly correlates with the quality of the generated images. Large, high-resolution, and diverse datasets are crucial for achieving realistic results. Researchers have focused on creating and curating such datasets specifically tailored for training GANs and other generative models, demonstrating the collaborative nature of the field, emphasizing dataset quality as much as architectural innovations.

The evolution of StyleGAN provides valuable insights for researchers and practitioners alike. It highlights the iterative nature of research in generative models, where incremental improvements and refinements contribute to significant leaps in performance and quality. The journey from StyleGAN to its successors showcased not only the power of deep learning but also the importance of persistent refinement, adaptation, and a focus on addressing subtle limitations.

Understanding the evolution of StyleGAN provides a compelling case study in the field. It's more than just a series of architectural improvements; it's a testament to the iterative nature of research and development in generative models. The journey from the initial StyleGAN architecture to the refined models that followed demonstrates the ongoing refinement and progress in tackling the inherent challenges within GAN training and pushing the boundaries of realistic image synthesis. This progression is essential for anyone seeking to understand and utilize the latest advancements in image generation. Each step, from addressing artifacts to enhancing stability, represents a significant contribution to our understanding and the ongoing efforts to create ever more sophisticated and realistic generative models. The legacy of StyleGAN extends far beyond its specific architectures; it represents a paradigm shift in how we approach and conceptualize image generation.

5.2 Improving GAN Stability

The notorious instability of Generative Adversarial Networks (GANs) is a persistent hurdle for researchers and practitioners alike. Mode collapse, where the generator produces limited variations, and vanishing gradients, hindering the learning process, are common issues. Addressing these challenges is crucial for reliable and creative image generation. Careful consideration of hyperparameters, particularly learning rates and batch sizes, is paramount. Experimentation is key, as optimal settings vary significantly depending on the dataset and network architecture. Small, incremental adjustments, rather than drastic changes, often yield better results. This iterative approach allows for a more nuanced understanding of how different parameters affect the training dynamics. Don't underestimate the power of meticulous hyperparameter tuning; it's a foundational aspect of achieving stable GAN training.

Regular monitoring of the training process is essential. Visualizing the generated images throughout training allows for early detection of mode collapse or other irregularities. Using tools that track key metrics, such as the generator and discriminator losses, offers invaluable insights into the network's behavior. Significant deviations from expected trends might indicate problems requiring intervention. For instance, a consistently low discriminator loss suggests the generator is easily fooling the discriminator, potentially leading to mode collapse. Conversely, a high discriminator loss might signal a struggle for the discriminator to learn effectively, potentially causing the generator's training to stall. Active observation enables proactive adjustments.

Beyond hyperparameter tuning and monitoring, architectural choices significantly influence stability. Consider incorporating techniques like spectral normalization, which stabilizes the discriminator by constraining its weight norms, preventing excessively large gradients. Alternatively, exploring different GAN architectures beyond the basic GAN structure, such as improved versions like Deep Convolutional GANs

(DCGANs) or those incorporating attention mechanisms, could improve stability and the quality of results. This careful choice of architecture can greatly mitigate the risks of instability.

Data quality plays a crucial role. Insufficient or low-quality data leads to unstable training. Ensuring a diverse and representative dataset is fundamental. Consider techniques like data augmentation, carefully expanding your dataset to include variations of existing images. This augmentation can prevent overfitting and enhance the model's ability to generate diverse outputs, indirectly contributing to increased stability. Furthermore, preprocessing your data, such as normalization and resizing, also enhances the training process, reducing noise and allowing the network to focus on relevant features.

Finally, don't disregard the significance of training strategies. Techniques like label smoothing, which reduces the confidence of the discriminator's predictions, or employing different loss functions beyond the standard minimax loss, can significantly impact the training process. Experimenting with different optimizers like Adam or RMSprop, adjusting their hyperparameters, might also prove beneficial. Furthermore, consider using techniques like gradient penalty or Wasserstein distance to improve training stability and the quality of generated samples. The pursuit of stable GAN training necessitates experimentation and adaptation, and continuous exploration is crucial for success. Remember that GAN training is an iterative process, and finding the right combination of techniques often involves trial and error.

6 Photorealistic Image Synthesis

Achieving photorealism in image synthesis requires a multi-pronged approach, going beyond simply increasing resolution. High-resolution generation, while crucial, isn't sufficient on its own; the resulting images often suffer from artifacts that betray their synthetic nature. These artifacts, ranging from subtle inconsistencies in textures to more glaring blurring or unnatural lighting, can significantly detract from the overall realism. Successfully addressing these issues requires a deep understanding of both the underlying generative models and the nuances of visual perception.

One critical aspect is the careful selection and preprocessing of training data. A dataset rich in diverse, high-quality images is paramount. However, simply amassing a large dataset isn't enough; the data must be carefully curated to ensure consistency and reduce bias. This often involves removing images with artifacts, noise, or inconsistencies that could negatively affect the model's learning process. Furthermore, careful consideration of data representation, including color spaces and normalization techniques, can significantly impact the quality of the generated images. For instance, utilizing perceptual loss functions during training can guide the model towards generating images that are perceptually more realistic, rather than just mathematically close to the training data.

Data augmentation, a technique used to artificially expand the training dataset, plays a crucial role in improving the robustness and generalizability of the model. Techniques like random cropping, flipping, and color jittering can introduce variability into the training data, helping the model learn more robust features and preventing overfitting. More advanced techniques, such as style transfer or texture synthesis applied to the training data, can further enhance the model's ability to generate diverse and realistic images. The key here lies in carefully choosing augmentation methods that preserve the overall realism and integrity of the data while introducing sufficient variation. Overly aggressive augmentation can lead to undesirable artifacts in the generated images, effectively undoing the benefits.

Addressing artifacts directly requires a combination of techniques, often interwoven within the model architecture itself. Improved loss functions, going beyond simple pixel-wise comparisons, are essential. Perceptual loss functions, which measure differences in higher-level image features, are far more effective at capturing subtle discrepancies that might be missed by simpler metrics. Furthermore, regularization techniques, aimed at preventing overfitting and promoting smoother, more natural-looking images, are crucial. These techniques often involve adding constraints to the model's parameters during training, thus subtly guiding it towards producing images with fewer artifacts.

Beyond these technical considerations, the choice of generative model architecture significantly influences the realism of the resulting images. While GANs have historically been at the forefront of photorealistic image synthesis, recent advancements in diffusion models and other architectures are closing the gap, and in some cases even exceeding GAN performance. The choice depends on various factors, including the specific application, the available computational resources, and the desired level of control over the generation process. Each model architecture comes with its own set of strengths and weaknesses regarding the generation of fine details, handling of complex textures, and control over the overall style of the generated images. Researching and comparing various architectures is crucial for choosing the most suitable method for a specific task. Continuous monitoring of model output and iterative refinement are crucial for identifying and correcting issues as the training process progresses. Careful analysis can reveal recurring issues specific to the selected model and dataset, enabling tailored solutions to maximize realism.

6.1 High-Resolution Image Generation

Generating high-resolution images presents unique challenges. While lower-resolution methods often rely on simpler algorithms, achieving photorealism necessitates handling significantly more data and detail. This requires advanced architectures and meticulous attention to detail, going beyond simple pixel manipulation. We're not just increasing pixel count; we're striving for realism that mimics the intricacies of natural scenes or meticulously crafted artwork.

Consider the sheer volume of information within a high-resolution image. The number of pixels explodes, and with them, the complexity of relationships between those pixels. A simple texture becomes a complex interplay of light, shadow, and subtle variations in color and tone. This necessitates sophisticated models capable of capturing and reproducing these nuances. Unlike simpler methods, we need models that understand not just individual pixels but the holistic structure and context of the entire scene.

Training these models demands significant computational resources. The datasets required to capture the necessary diversity and fidelity are immense, often comprising millions of high-resolution images. This presents a hurdle in terms of both storage and processing power, necessitating optimized training techniques and potentially specialized hardware. Simply increasing the size of a lower-resolution model won't suffice; the underlying architecture needs to be designed for scalability and efficiency.

One major obstacle in high-resolution image generation is the appearance of artifacts. These can manifest as blurring, distortions, or inconsistencies in texture, undermining the photorealism we aim for. These imperfections arise from limitations in the model's ability to capture and reproduce complex relationships within the image. Effective mitigation strategies often involve careful model design, advanced regularization techniques, and potentially post-processing steps to refine the generated output.

Data augmentation plays a critical role in training high-resolution models. Standard augmentations like rotation or cropping are less effective at the high-resolution level. We need to focus on transformations that maintain the structural integrity of complex scenes. For example, carefully constructed patches of high-resolution images can be used to synthesize new regions, while techniques like super-resolution can be employed to further enhance detail and realism, ensuring a consistent level of visual quality across the entire image. Moreover, the types of augmentations must be carefully considered to avoid introducing unintended biases or distortions.

Furthermore, the evaluation of high-resolution image generation presents a unique challenge. Traditional metrics like PSNR (Peak Signal-to-Noise Ratio) are often inadequate as they don't capture perceptual qualities. Instead, we often rely on subjective assessments by human experts, alongside more sophisticated metrics that attempt to measure perceptual similarity and the fidelity of fine details. This necessitates careful consideration of evaluation methodologies, and often a blend of quantitative and qualitative measures. The sheer size of the data makes automated evaluation challenging, but it is crucial for tracking the progress and success of the models.

Finally, the applications of high-resolution image generation extend beyond mere visual appeal. In scientific visualization, for instance, generating high-resolution representations of complex datasets allows for detailed analysis and exploration, uncovering patterns and insights that might otherwise be missed. Similarly, high-resolution image synthesis enhances capabilities in fields such as medical imaging, offering the potential for improved diagnostics and treatment planning. The ability to generate realistic and detailed images is no longer just an aesthetic goal; it's a crucial tool for a range of critical applications. Understanding the challenges, and employing appropriate strategies, is paramount in realizing its full potential.

6.2 Addressing Artifacts

High-resolution image generation, while incredibly powerful, often produces artifacts—undesirable features that detract from the realism and visual appeal of the synthetic image. These imperfections can manifest in various ways, from subtle inconsistencies in texture and shading to more glaring anomalies like blurry regions or unnatural color banding. Understanding these artifacts and employing effective mitigation strategies is crucial for achieving truly photorealistic results.

One common artifact is "checkerboarding," where distinct square patterns appear in the generated image, particularly noticeable in areas with sharp contrast or fine details. This is frequently observed in GAN-based models and often stems from limitations in the model's ability to capture high-frequency information. Addressing this involves careful model architecture design, employing techniques like improved upsampling methods or incorporating perceptual losses that penalize checkerboarding patterns during training. Experimentation with different hyperparameters, such as the learning rate and batch size, can also yield positive results. Moreover, a careful selection of the dataset, ensuring sufficient diversity and high-resolution images, is fundamental to achieving better results.

Blurriness, another pervasive issue, is often caused by insufficient model capacity or training data. The generated image lacks the crispness and detail of a real photograph, appearing soft and indistinct, especially in areas with intricate textures or fine lines. To tackle this, consider using larger models with more parameters, and the use of higher resolution training images is paramount. Implementing techniques such as super-resolution models during or after image generation can help refine detail and sharpen the final output. Also, incorporating loss functions that specifically reward sharp, detailed output may considerably improve the final results.

Color banding, characterized by abrupt transitions between distinct color shades instead of smooth gradients, significantly hampers the realism of an image. This artifact usually arises from a limited color palette used in the generation process or from numerical limitations within the model itself. Increasing the bit depth of the output image can reduce color banding, which will lead to a much smoother appearance. Advanced techniques like dithering, which strategically introduces noise to simulate more colors, can also help to soften the transitions and make the image more visually appealing. The usage of higher-quality training data that has a more nuanced color range is important, as this provides the model with a broader range of color transitions to learn from.

Beyond these common artifacts, other irregularities can emerge, including texture inconsistencies, unnatural lighting, and distorted shapes.

Identifying the specific source of these artifacts requires a keen eye for detail and often demands careful examination of the model architecture and training process. This means paying very close attention to dataset composition, model hyperparameters, and loss function design is essential for identifying the root cause and applying the correct adjustments. Addressing artifacts is an iterative process involving meticulous experimentation and careful analysis. Begin by identifying the dominant artifacts in your generated images. Then, systematically explore various mitigation techniques, carefully assessing their impact on the overall image quality. Don't be afraid to combine multiple approaches; a multifaceted strategy often yields the best results. Remember, the goal is not simply to eliminate all artifacts but to create images that are visually compelling and consistent with the desired level of realism. The pursuit of photorealism is a continuous journey of improvement and refinement. As you gain experience, you'll develop a better intuition for identifying and addressing these challenges, leading to more successful results in your image generation projects.

6.3 Data Augmentation Strategies

Data augmentation is crucial for training robust and high-quality image generation models, particularly when aiming for photorealism. Insufficient training data often leads to models that overfit, generating images that look similar to the training set but lack generalization. Augmentation combats this by artificially expanding the dataset, introducing variations that help the model learn more diverse and realistic features. Simple transformations like flipping images horizontally or vertically can significantly boost performance. These basic methods, however, often prove insufficient for truly photorealistic results.

More sophisticated techniques become necessary. Consider color jittering, where small random changes are introduced to the hue, saturation, and brightness of images. This mimics real-world variations in lighting and conditions, forcing the model to learn more robust representations of colors and textures. Similarly, random cropping and resizing can create variations in composition and scale, improving the model's ability to recognize objects regardless of their size or placement within the frame. These augmentations are relatively straightforward to implement and yield significant benefits, especially when combined.

Beyond these foundational methods lies the potential of more complex transformations. Imagine applying various noise patterns, from subtle grain to more pronounced distortions. This simulates imperfections inherent in real-world photographs, enriching the model's training data with a wider spectrum of visual irregularities. Careful consideration should be given to the type and intensity of noise introduced, to avoid introducing artifacts that could negatively impact the final generated image quality. The key is to introduce realistic noise, not artificial noise that detracts from the model's performance.

The power of data augmentation extends to more advanced techniques, such as incorporating geometric transformations. These could include rotations, shears, and perspective warps, simulating different viewpoints and orientations. However, these can be computationally expensive, necessitating careful balancing between the benefits of such augmentations and their implementation cost. Effective strategy often involves starting with simpler augmentations and progressively adding more complex techniques, evaluating the impact at each stage. It's vital to remember that not all augmentations are equally effective; experimentation and careful evaluation are essential.

Furthermore, the generation of synthetic data offers another powerful augmentation method. Using existing models or simpler techniques, you can create new images that resemble your target dataset, filling in gaps and increasing data diversity. However, the quality of synthetic data significantly impacts the outcome, and the use of low-quality synthetic images can even hinder the training process. Careful selection and validation of synthetic data are therefore critical steps in this augmentation strategy. Remember to validate synthetic data rigorously, ensuring it doesn't introduce unwanted biases or artifacts into your dataset.

Consider also the use of adversarial examples. These are subtly modified images designed to trick the model into misclassifying or generating unrealistic outputs. By incorporating these adversarial examples into the training dataset, we make the model more robust to subtle variations and perturbations, leading to improved generalization and resistance to adversarial attacks. Generating these, however, requires sophisticated techniques and a deep understanding of the model's inner workings, making it a more advanced augmentation strategy. This can be incredibly useful for refining the model's ability to handle edge cases.

The process of data augmentation isn't just about applying random transformations; it's about carefully selecting and combining various techniques to address specific shortcomings in your dataset. For example, if your dataset lacks diversity in lighting conditions, augmentations focusing on color jittering and simulated shadows become particularly important. Similarly, if your images are primarily focused on a narrow range of viewpoints, then incorporating geometric transformations like rotations becomes critical. Ultimately, the best strategy will depend on the specific characteristics of your dataset and the desired output.

Finally, remember that data augmentation is an iterative process. You should monitor the performance of your model closely throughout the training process, analyzing the generated images and adjusting your augmentation strategy as needed. This could involve adjusting the parameters of existing augmentations, adding new transformations, or removing ineffective ones. The key is to strike a balance between augmenting the dataset sufficiently to improve model performance and avoiding the introduction of artifacts or unrealistic features. Effective data augmentation is a crucial step towards achieving truly photorealistic image generation, demanding a careful and considered approach.

7 Creative Applications & Future Trends

The application of image generation extends far beyond mere aesthetics; it's revolutionizing various fields. Artists are utilizing these tools to create breathtakingly original pieces, pushing the boundaries of traditional mediums and exploring new creative avenues. Imagine an artist seamlessly blending photographic realism with fantastical elements, a process once painstakingly slow, now achievable with remarkable speed and precision. This accessibility democratizes artistic creation, empowering individuals with limited technical skills to realize their creative visions. The impact on the art world is profound, challenging conventional notions of authorship and originality, and sparking exciting dialogues about the nature of art itself.

Game developers are harnessing the power of image generation to build immersive and visually stunning worlds. Procedural generation, for instance, allows for the creation of vast, unique landscapes, eliminating the need for painstaking manual design of every tree, rock, and building. This efficiency allows for a focus on gameplay mechanics and storytelling, resulting in richer and more engaging gaming experiences. Think about the potential: dynamic environments that change with the seasons, procedurally generated characters with unique appearances, and vast, explorable worlds limited only by the imagination of the developers. The efficiency gains are substantial, paving the way for more complex and detailed game worlds.

Beyond entertainment, image generation is transforming scientific visualization. Researchers across disciplines are using these tools to interpret complex datasets, visualizing abstract concepts in an intuitive and accessible manner. Imagine visualizing the intricate structures of proteins, simulating the spread of diseases, or modeling climate change impacts—all with incredible visual clarity and detail. This allows for a deeper understanding of complex phenomena, fostering collaboration and accelerating scientific discovery. These advancements offer a transformative way to bridge the gap between complex data and intuitive understanding.

However, the widespread adoption of image generation technologies necessitates careful consideration of ethical implications. Biases embedded in training datasets can perpetuate and amplify existing societal inequalities, resulting in generated images that reflect and reinforce harmful stereotypes. The potential for misuse, including the creation of deepfakes and the spread of misinformation, requires vigilance and proactive measures to mitigate these risks. Responsible development and deployment of these technologies are crucial to prevent unintended consequences and ensure equitable access to their benefits. Furthermore, copyright and intellectual property issues surrounding generated images remain complex and require further legal clarity.

Looking ahead, the future of image generation is brimming with exciting possibilities. Improvements in model efficiency and realism are continuous, leading to the creation of even more lifelike and detailed images. The integration of other advanced technologies, such as augmented and virtual reality, promises to create even more immersive and interactive experiences. We might even see the emergence of completely new artistic mediums and creative processes, enabled by the capabilities of advanced image generation systems. The trajectory is toward increasingly sophisticated algorithms capable of understanding and responding to complex prompts and creative instructions, opening unprecedented avenues for artistic expression and technological innovation. The field is ripe for further exploration and groundbreaking discoveries.

7.1 Art & Design Applications

The ability to generate images using algorithms opens up a plethora of possibilities for artists and designers. Imagine creating unique textures for fabrics, generating intricate patterns for wallpapers, or designing bespoke logos – all without the constraints of traditional design tools. This isn't science fiction; it's the present reality, and the possibilities are continually expanding. This transformative power empowers creatives to explore uncharted aesthetic territories and boost their productivity.

Beyond the straightforward applications, generative models offer unique creative avenues. For example, consider using a generative model to create a series of variations on a single design element, allowing for rapid prototyping and experimentation with different visual styles. This iterative approach facilitates a more dynamic and less constrained design process, allowing for rapid exploration of diverse ideas. By leveraging the strengths of algorithms, designers can overcome limitations and break through creative barriers.

The collaboration between human creativity and algorithmic power is particularly compelling. Instead of being a replacement for the human artist, generative models act as a powerful extension of creative capabilities. The designer can input their vision, guiding the algorithm to refine and evolve its output. This iterative interaction, this back-and-forth between artist and algorithm, creates a synergistic relationship that produces unique and striking results, a true hybrid creation.

One could envision a workflow where a designer starts with a rough sketch, feeding it into a generative model to generate variations on that sketch, each with slightly different textures, colors, and styles. The designer then selects the most promising variations, making further refinements and edits until the desired aesthetic is achieved. This approach not only speeds up the design process but also fosters unexpected discoveries, leading to innovative designs that would have been difficult, or perhaps even impossible, to conceive using traditional methods alone.

This isn't limited to static images; consider animation. Generative models can assist in creating unique animations, from subtle background elements to complex character movements. Imagine designing an animated film where the backgrounds are algorithmically generated to perfectly complement each scene, creating a visual harmony that's difficult to achieve manually. This allows for much more efficient workflow and also for more creative explorations of visual storytelling.

Moreover, the use of generative models in design software isn't confined to image generation itself. Imagine using generative models to create personalized user interfaces, adapting the design to suit individual preferences. Imagine creating generative design tools that offer suggestions and create diverse iterations based on user input, a design assistant that's always learning and adapting to the designer's style and preferences. The opportunities for improvement in user experience through these generative models are numerous and constantly evolving.

The applications extend beyond individual creation. Generative models can also be employed for collaborative design projects. Imagine a team of designers working on a single project, each contributing variations and iterations using generative models, and then combining the best aspects of each to create a cohesive and powerful final product. This collaborative process allows for a broader range of ideas and a higher level of creative exploration than traditional methods can offer.

The ethical considerations are also important, however. While generative models offer incredible potential, it's crucial to be mindful of copyright implications and potential biases in the data used to train these models. Using generative models responsibly requires an understanding of these concerns, and a commitment to ethical practices in both the creation and application of these powerful technologies.

Ultimately, the integration of generative models into art and design is revolutionizing the creative process, not by replacing human artists, but by empowering them with unprecedented tools and opportunities. The ability to explore endless possibilities, to experiment freely, and to collaborate with algorithms, opens up new dimensions of creative expression and expands the horizons of visual art and design. This is a collaborative partnership, leading to truly novel and exciting outcomes. The future of art and design is generative, and the creative possibilities are limitless.

7.2 Game Development

Game development presents a unique and exciting application for image generation techniques. Beyond simple sprite creation, these techniques offer the potential to revolutionize aspects like environment design, character modeling, and even procedural content generation, significantly reducing development time and expanding creative possibilities. Imagine crafting vast, intricate game worlds with minimal manual effort, populating them with unique characters and items, all powered by the algorithms we've explored throughout this book.

The integration of generative models, especially GANs and VAEs, opens doors to previously unimaginable levels of detail and variety. Consider the time-consuming process of creating unique textures for every item in a game; generative models can significantly accelerate this. By training a GAN on a dataset of existing textures, developers can generate a near-infinite supply of variations, each subtly different yet cohesive with the overall game style. This allows for a level of consistency and artistic detail that's simply not achievable through traditional manual methods. Character creation benefits similarly. Instead of meticulously crafting each character model from scratch, developers can use generative models to create a diverse range of characters, complete with unique facial features, clothing, and even animations. This streamlines the process dramatically, freeing up artists to focus on refining key characters or creating high-quality assets for specific scenes. Moreover, VAEs can help in generating variations of existing character models, providing alternate outfits, expressions, and poses, enhancing the gameplay experience with richer character representation.

Procedural generation is another area where image generation shines. Using algorithms, you can automatically generate levels, landscapes, or even entire game worlds, making it possible to create massive, engaging games with diverse and unpredictable content. Imagine a game world that's never truly the same twice, because each playthrough unfolds in a procedurally generated environment, populated with unique items and creatures, presenting players with a fresh experience each time. This dramatically increases replayability.

However, seamlessly integrating generative models into a game engine requires careful consideration. The computational cost of generating images in real-time can be substantial, particularly with high-resolution models. Therefore, efficient techniques are crucial, often involving pre-generation of assets and smart caching strategies. Developers might pre-generate a library of assets—textures, character models, environments—and then utilize those assets within the game, calling upon them as needed, rather than generating everything on the fly.

Furthermore, controlling the style and consistency of the generated assets is paramount. While the ability to create a vast array of variations is advantageous, maintaining a consistent visual style across the game is equally important. This requires careful training of the generative models and potentially incorporating additional constraints into the generation process, ensuring the output aligns with the desired aesthetic. This involves using techniques like style transfer or carefully crafting the training data to reflect the intended artistic direction.

Beyond visual elements, image generation can be leveraged to create dynamic game environments, ones that change and evolve organically over time. Imagine a forest where trees grow and change seasons, or a city where buildings are constructed and demolished based on in-game events. This level of dynamic detail vastly enhances the realism and immersion, transforming the game experience from a static world into a living, breathing ecosystem. It's no longer a matter of creating a fixed environment; instead, the environment itself is part of the gameplay experience.

The potential uses extend even further. Generative models can aid in creating realistic lighting effects, generating unique soundscapes based on environmental context, and even crafting realistic character animations. These applications, combined with the ability to generate varied and unique content, make image generation an indispensable tool for modern game development. But remember, success hinges not just on implementing these techniques but on intelligently integrating them into the game's overall design and gameplay experience, ensuring they enhance, not detract, from the player's journey. Mastering this integration is the true key to unlocking the transformative power of image generation in game development.

7.3 Scientific Visualization

Scientific visualization transforms complex datasets into visually intuitive representations, making abstract information readily comprehensible. Imagine a researcher struggling to interpret terabytes of climate data—numbers and graphs alone offer limited insight. However, a well-designed visualization, perhaps a dynamic globe displaying temperature changes over time, paints a clear, compelling picture, revealing trends and patterns instantly. This is the power of scientific visualization: translating the quantitative into the qualitative, the abstract into the concrete.

This transformative power extends far beyond climate science. Medical imaging, for instance, relies heavily on visualization. From X-rays and MRIs to advanced 3D models of organs and tissues, visual representations allow doctors to diagnose diseases, plan surgeries, and monitor treatment progress with a precision previously unimaginable. The visual clarity facilitates better decision-making and ultimately, improved patient outcomes. Consider the intricate network of blood vessels in the human brain, too complex to understand fully from static data alone. A 3D visualization, however, allows for detailed examination, revealing blockages or anomalies that might otherwise remain hidden.

The creation of effective scientific visualizations requires a careful blend of art and science. While technical expertise in data processing and software is essential, artistic sensibilities are equally crucial. The visual representation must not only be accurate but also aesthetically pleasing and engaging, capturing the viewer's attention and guiding their understanding. It's a delicate balance: simplifying complexity without sacrificing crucial details. Choosing the right color palettes, for example, can significantly impact the interpretation of the data; inappropriate choices might lead to misinterpretations and skewed conclusions. Understanding the principles of visual perception is vital, ensuring the visualization effectively communicates the intended message to its intended audience.

Moreover, interactive visualizations offer enhanced engagement and exploration. Instead of static images, viewers can manipulate the representation, zooming in on specific areas, rotating 3D models, or filtering data to examine specific aspects. This dynamic interaction encourages deeper engagement and a more thorough understanding of the underlying data. Interactive tools empower the user to explore complex relationships, discover hidden patterns, and develop a more intuitive grasp of scientific concepts. Consider a simulation of a fluid flow, allowing the user to adjust parameters and observe the effects in real-time. This kind of dynamic interaction offers unparalleled insight, facilitating hypothesis testing and problem-solving.

Beyond the technical aspects, consider the ethical implications. Scientific visualizations are not merely passive displays of information; they are powerful tools that can shape our understanding of the world, influencing public opinion and policy decisions. It is crucial that these representations be created responsibly, avoiding bias, ensuring accuracy, and providing context to prevent misinterpretations. Overly simplistic or misleading visualizations can have significant consequences, leading to flawed conclusions and uninformed choices. Transparency in the data processing and visualization methods is paramount, allowing users to critically assess the validity and reliability of the presented information.

Ultimately, mastering scientific visualization is about more than just creating pretty pictures. It's about communicating complex scientific concepts effectively, enabling collaboration, and advancing knowledge across various fields. Through careful consideration of the data, the target audience, and the technical and aesthetic choices, researchers and scientists can use the power of visualization to unlock deeper insights, facilitate discoveries, and ultimately, improve our world. It's a field demanding both technical proficiency and artistic sensitivity, a rewarding pursuit for those seeking to bridge the gap between the complexities of scientific data and the clarity of visual understanding.

7.4 Ethical Considerations

The power to create realistic images from nothing has profound ethical implications. We must consider the potential for misuse, from the creation of deepfakes used for malicious purposes to the spread of misinformation disguised as authentic visuals. This technology, while incredibly powerful for creative endeavors and scientific advancement, also presents a significant threat to trust and social stability. Navigating this complex landscape requires careful consideration and proactive measures. The ease with which convincing yet false imagery can be generated necessitates the development of robust detection methods, promoting media literacy and critical thinking among the general public. Ignoring this aspect is irresponsible; proactively addressing these challenges is crucial for ensuring responsible technological development.

Developing and implementing ethical guidelines is paramount. These guidelines should cover data privacy, intellectual property rights, and the potential for algorithmic bias. The datasets used to train these models often contain biases reflecting societal inequalities; these biases can be amplified and perpetuated by the generated images, leading to the reinforcement of harmful stereotypes. This requires rigorous auditing of datasets and the implementation of fairness-enhancing techniques within the algorithms themselves. Moreover, the potential for copyright infringement when training models on copyrighted material needs thorough legal and ethical examination. A clear framework is necessary to balance technological advancement with the protection of creators' rights.

Consider the impact on artists and their livelihoods. The accessibility of image generation tools raises concerns about the potential displacement of human artists, particularly those working in fields that are readily automatable. However, this technology also presents opportunities for collaboration and creative augmentation. Rather than viewing it as a purely disruptive force, we can explore ways to integrate it into artistic workflows, enhancing creative processes and enabling new forms of expression. Education and training initiatives are crucial in preparing artists to adapt to this evolving landscape and utilize these tools effectively in their practices. This requires fostering dialogue and finding ways to leverage technology in a way that complements, not supplants, human creativity.

Transparency is a cornerstone of ethical image generation. Users should be aware when they are interacting with synthetic images, to avoid deception and misinformation. Watermarking techniques and other forms of provenance tracking are being developed, but these methods require careful evaluation to avoid compromising the quality or usability of the images. A balance must be struck between providing clear indicators of synthetic origin and avoiding overly intrusive or cumbersome processes that hinder the technology's accessibility. Open communication regarding the limitations and potential biases of these models is also essential. This encourages a critical approach to assessing the authenticity and reliability of imagery, especially in contexts where misinformation can have serious consequences.

The future of ethical image generation depends on a multi-faceted approach. This includes rigorous research into bias detection and mitigation, the development of robust verification methods, the establishment of clear legal frameworks, and the implementation of industry-wide ethical standards. Educational programs should be implemented to increase media literacy and critical thinking skills, helping individuals discern between genuine and synthetic imagery. By fostering collaboration between technologists, artists, legal experts, and policymakers, we can harness the creative potential of image generation while mitigating its potential risks. This necessitates ongoing dialogue, adaptation, and a shared commitment to utilizing this technology responsibly, ensuring a future where it serves humanity rather than undermines it.

7.5 Emerging Technologies

Newer methods are constantly emerging, pushing the boundaries of what's possible in image generation. Consider, for instance, the integration of physics-based simulations into generative models. Instead of relying solely on learned patterns from existing datasets, these models incorporate principles of physics, allowing for the creation of images that realistically simulate natural phenomena like fluid dynamics or light scattering. This opens doors to more accurate and visually compelling depictions of complex scenarios, something previously difficult to achieve without considerable manual intervention. The implications are vast, from realistic simulations of weather patterns to virtual environments that closely mirror the real world.

Neural radiance fields (NeRFs) represent another significant advancement. These models learn a continuous representation of a scene from multiple 2D images, enabling the generation of novel views of that scene. Imagine reconstructing a historical site from a handful of photographs, or creating interactive 3D models from simple video footage. This capability transforms how we interact with and visualize information, moving beyond static images to immersive experiences. The technical hurdles are significant, demanding high computational power and carefully designed algorithms, but the potential impact on applications like virtual tourism, architectural visualization, and filmmaking is undeniable. Furthermore, the rise of generative models that operate directly on 3D data is revolutionizing the field. Instead of generating 2D images, these models create three-dimensional objects or scenes, offering far greater flexibility and control. This is crucial for applications in game development, where detailed and realistic 3D models are crucial for building immersive gaming environments. Beyond gaming, this technology is transforming various fields such as product design, allowing designers to iteratively refine 3D models with minimal effort. The capacity to quickly explore numerous design variations can speed up the design process and lead to innovative designs that might otherwise be missed.

The increasing accessibility of these tools is also important. While many advanced techniques still require specialized expertise and computational resources, user-friendly interfaces and cloud-based platforms are making powerful generative tools accessible to a broader range of users. This democratization of image generation technology is fostering creativity and innovation across various fields, from digital art to scientific visualization. This empowers individuals to explore their artistic visions and contribute to scientific discoveries without having to possess in-depth technical expertise.

However, it's crucial to acknowledge the challenges. Many emerging technologies struggle with issues of scalability, computational cost, and the potential for generating biased or unrealistic outputs. Researchers are actively working on addressing these limitations, exploring more efficient algorithms and methods for training models on larger and more diverse datasets. The ongoing pursuit of ethical considerations remains paramount, and addressing bias and promoting responsible use of generative models will be essential to ensuring the beneficial use of this powerful technology. Without these safeguards, the potential risks of misinformation and misuse must be carefully mitigated.

Finally, the future of image generation will likely be shaped by the synergy between different approaches. We can expect to see more sophisticated hybrid models that combine the strengths of deep learning with physics-based simulations or other techniques. This integrated approach would leverage the power of data-driven learning with the accuracy and consistency of physically informed models, resulting in even more realistic and believable images. The combined approaches could pave the way for entirely new creative possibilities and push the boundaries of what's achievable in image synthesis far beyond our current capabilities. The journey continues, leading towards a more realistic and dynamic image generation landscape.

www.ingramcontent.com/pod-product-compliance
Lightning Source LLC
LaVergne TN
LVHW022015200926
848481LV00007B/2406
* 9 7 9 8 3 4 8 5 2 5 9 9 6 *